This is YOUR BODY

By Gyles Brandreth

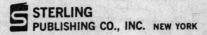
STERLING PUBLISHING CO., INC. NEW YORK

Other Books by Gyles Brandreth
The Biggest Tongue Twister Book in the World
Brain-Teasers and Mind-Benders
A Joke-A-Day Book
Pranks, Tricks and Practical Jokes

Special School Book Fair Edition

Copyright © 1979 by Sterling Publishing Co., Inc.
Two Park Avenue, New York, N.Y. 10016
Original edition published in Great Britain
in 1978 under the title "Project: The Human Body"
by Corgi-Carousel Books, a division of Transworld
Publishers Ltd., copyright © 1978 by Gyles
Brandreth, illustration copyright © 1978 by
Transworld Publishers Ltd.
Manufactured in the United States of America
All rights reserved
Library of Congress Catalog Card No.: 79-65078
Sterling ISBN 0-8069-3112-4 Trade
3113-2 Library

CONTENTS

Introduction	4
All in the Body	5
Blood, Blood, Glorious Blood!	8
Chatterbox	18
Deep Breathing	22
Ear! Ear!	27
Food for Thought	32
Germ Warfare	36
Hair Raising	39
Into the Intestines	43
Joints and Bones	47
Kidney Korner	55
Left-Handed	57
Muscle Power	61
Nose Knows	67
Optic Organs	71
Perception	77
Quick Thinking	84
Records	89
Skin Tight	92
Teeth Time	95
Ugh!	99
Vocabulary	102
What Goes Where?	106
X-Rays	109
You've Got a Nerve	111
ZZZzzzzzz	117
Index	120

This is a book I should have read when I was young—but it wasn't written then, so I had to wait until I was at the university before I discovered how the human body is made and how it works. It was never as much fun for me learning it all the hard way as it will be for you reading this excellent text, understanding the diagrams and laughing at the cartoons. Even if you don't want to be a doctor, nurse, or other health worker, it is still a good idea to know about your own body. It's the only one you've got, so learn how it works, and then you'll know how to look after it.

GYLES BRANDRETH

ALL IN THE BODY

*What are little girls made of?
Sugar and spice
And all things nice,
That's what little girls
Are made of.*

*What are little boys made of?
Frogs and snails
And puppy dogs' tails,
That's what little boys
Are made of.*

"RUBBISH!" you say and quite rightly so. But what are we all made of?

Well, we're made up—each one of us—of skin and bones, of muscles and nerves, of heart and brain, and of lots of other things besides.

In an adult there is:

> enough fat to make two dozen big bars of soap—
> enough sulfur to kill all the fleas on a dog—
>
> enough lime to whitewash a chicken coop—
> enough carbon to make the lead for 9,000 pencils—
> enough iron to make a solid nail—
> enough water to fill a 2-gallon jug.

Yes, but what makes up the skin and the bones, the muscles and the nerves, the heart and the brain, the fat and the sulfur and the lime and the carbon and the iron and the water and everything else? The answer: **Cells.**

Our bodies are made of millions of tiny living cells. There are more than 50,000,000,000,000 cells in one human body. Most of them are so small that you could fit 100,000 of them onto the head of a pin!

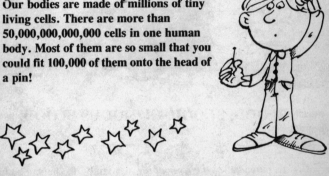

Each part of your body that this book describes is made up of a mass of cells. They come in all sorts of shapes and sizes. When they are grouped together they are called **Tissue**. Different parts of your body are made up of different kinds of body tissue. When different kinds of body tissue are organized to perform a particular kind of work the tissues form an **Organ**. When different organs work together to do a special job the group of organs is called a **System.**

The systems are made up of organs. The organs are made up of tissues. The tissues are made up of cells. And that's what this book and your body are really all about.

BLOOD, BLOOD, GLORIOUS BLOOD!

Blood is often called *the river of life*. Certainly, the human body would be lost without it! Its work is vital.

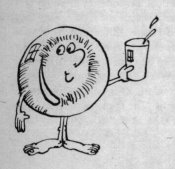

Blood supplies the cells of the body with water and nourishment.

Blood takes oxygen from the lungs and carbon dioxide to them.

Blood removes the waste products from the body's cells.

Blood carries heat from the hotter to the cooler regions.

Blood contains cells that fight disease and substances that can repair cut and bruised parts of the body.

Blood is made up of liquid and very tiny solid parts. The liquid part is called **plasma**. It is straw-colored and counts for more than half our blood (some 55 percent). The solid parts are called **red blood cells**, **white blood cells**, and **platelets**.

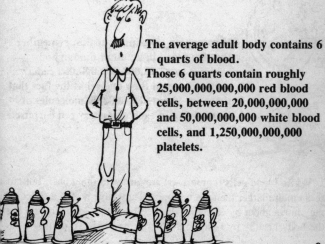

The average adult body contains 6 quarts of blood.
Those 6 quarts contain roughly 25,000,000,000,000 red blood cells, between 20,000,000,000 and 50,000,000,000 white blood cells, and 1,250,000,000,000 platelets.

Red blood cells (corpuscles) are round and pushed in on each side. They contain a substance made from iron called **hemoglobin**. Hemoglobin picks up oxygen from the lungs and delivers it to cells all over the body. Without oxygen, hemoglobin is dark blue, almost black. With oxygen it turns red, which is why the blood you see coming out of a cut is red.

If you enjoy amazing statistics, remember that the blood in an adult human body contains some 25,000,000,000,000 red corpuscles, and then marvel at the fact that there are about 280,000,000 molecules of hemoglobin in each and every red corpuscle!

White blood cells (corpuscles) are far fewer in number, but most of them are larger than the red ones. They have no fixed shape and they move about by changing shape. Their main job is to help keep the body healthy by fighting dangerous bacteria. To destroy a bacterium a white corpuscle moves over it and eats it up. It may die in the process and go (with the dead bacteria) to form pus. When large numbers of bacteria get into the blood, the body automatically produces more white blood cells to destroy the bacteria.

The **Platelets** form the part of the blood that makes it clot. If your blood didn't clot you could bleed to death with only the slightest cut. People who suffer from a condition called **hemophilia** have blood that does not clot quickly.

The **Plasma**, which makes up more than half the blood, is itself **91 percent water**. The rest of it is salts and proteins and what has been described as *the blood cargo*—the nourishment, waste products and so on that are being carried around the body.

The **Heart** is the wonderfully useful pump that moves our blood through the body. It is a muscle, about the size of your clenched fist, that contracts and relaxes about 70 times a minute. Each contraction and relaxation of the heart counts as one heartbeat.

**Your heart beats about 100,000 times a day.
In a year it beats about 36,000,000 times.
In an average lifetime a heart beats some 2,500,000,000,000 times!**

The heart is situated in the middle of the chest slightly to the left-hand side. It is divided into four chambers. The upper two chambers are called **auricles**, and the lower two chambers are called **ventricles**.

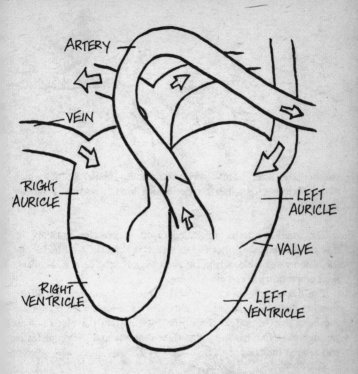

Each auricle (also called **atrium**) is connected with the ventricle below it by a valve that allows a flow **from** the auricle **to** the ventricle. When the right ventricle of the heart contracts, blood is forced into a large **artery** that leads to the **lungs**.

Arteries are the tubes that carry the blood **away** from the heart. It's in the lungs that the red corpuscles collect the oxygen and get rid of carbon dioxide. From the lungs, the blood goes through the **veins** that lead back to the heart.

Veins are the tubes that carry the blood **towards** the heart. Blood enters the left auricle and passes through the **valve** leading to the left ventricle. When the left ventricle contracts the blood flows into

another large artery. This artery branches into smaller arteries which branch into even smaller arteries which branch into even smaller arteries!

The smallest arteries are called **capillary arteries** and they are tiny tubes connecting arteries to veins. It is from the **capillaries** that the blood gives the nourishment and oxygen to the cells and collects the carbon dioxide and other waste.

The tiny capillaries connect with larger and larger veins. Blood flowing through these veins finally reaches a large vein that enters the heart's right auricle. From the right auricle the blood travels through the valve leading to the right ventricle and so completes its trip around the body.

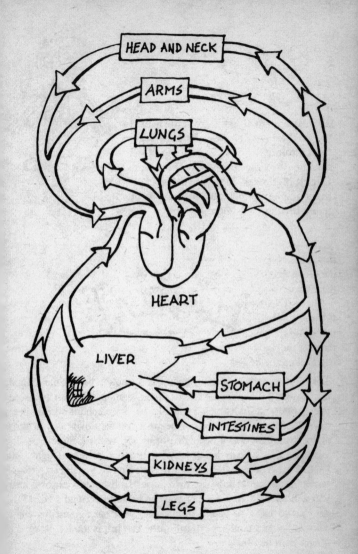

The round trip of blood from the heart to the farthest part of the body takes less than a minute. The circulation of blood through the lungs takes only six seconds. Over a lifetime of 70 years the heart pumps about 66,000,000,000 gallons of blood through the body.

1. How fast does your heart beat? As a rule the bigger the animal, the slower the heartbeat. For example, a mouse's heart beats between 600 and 700 times a minute, but an elephant's heart beats only 25 times a minute. Your heartbeat slows down as you grow older, so a baby has a heart that beats 130 times a minute but the average adult's heart beats only 70 times a minute. Count your heartbeats. You can do it either by listening to your heart through your chest or by feeling your own pulse. To listen to your own heart you will need a stethoscope or you can listen to a friend's heart (and get him to listen to yours) by pressing your ear against his chest. You will hear a double beat: **lub-dub**. The **lub** is louder, lower and longer than the **dub**.

To feel your own pulse, place a finger on your wrist like this:

When you feel your pulse, you will feel just one beat. Take a watch with a second hand and count up the number of beats in 60 seconds.

2. When you need lots of energy—for running or dancing or jumping—your heart will beat faster and pump more blood around your body. When it is really pushed the heart can pump five times the amount of blood that it normally pumps! Count your heartbeats after you have been resting for a while and then count them again after you have been running hard for a while. What's the difference?

CHATTERBOX

When you want to make a noise, this is how you do it:

1. You let air out of your lungs.
2. The air passes through your **windpipe**.
3. Your **vocal cords** are in your windpipe and they vibrate as the air pushes past them.
4. The vibrating vocal cords make **sound waves** in the air. Your lips and tongue and teeth can break up or squeeze the sound waves to make them into words.

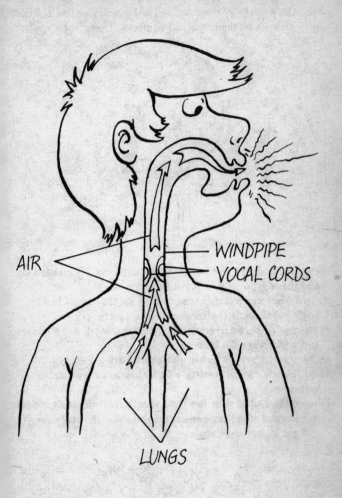

How high or how low the sound seems when you hear it is called **pitch**. The pitch of your voice, high and squeaky or low and deep, largely depends on the length and tightness of your vocal cords.

Both boys and girls are born with cords about 2½ inches long and these grow as you get older.

By two, they are 3 inches long and by six they are 4 inches.

A boy's vocal cords change more than a girl's.

At the age of 14, on average, a boy's vocal cords are 5 inches and a girl's are 4½ inches.

At 20, a boy's are 10 inches and a girl's are 6½ inches.

At 30, a man's vocal cords are 12 inches and a woman's are 8 inches.

Toward the end of your life your vocal cords will change again. But instead of getting longer, they will shorten and your voice will get higher once again.

Most animals can make sounds. Many can convey information in sounds, but none, not even the highly intelligent dolphin, use the voice in the way that human beings can.

The 4,000,000,000 or so human beings in the world speak some 2,000 different languages.

Mandarin Chinese, spoken by 550,000,000 people, is the world's most popular language.

English is the next most popular language. It is the mother tongue of 320,000,000 people.

Can you find out what are the world's 10 most spoken languages? And can you draw a Language Map of the World, showing in which parts of the world each of the different languages is most used?

DEEP BREATHING

The cells of the body need **oxygen**. You get oxygen from the a
In order to get the air from outside inside, we breathe in. This is hc
the system works:

1. To make you breathe in, the muscle under your lungs, which
   called the **diaphragm**, flattens down and your ribs swing up a
   out.
2. The pressure of the air outside your body is now greater than t
   pressure in your lungs, so air is drawn into your nose, goes dov
   your throat and windpipe, through a tube called the **trachea**, a
   then into your lungs.
3. The trachea divides into two parts. Each part is called a **bronch
   al tube** and each part enters a lung.

4. Each bronchial tube branches many times until the smallest branches are almost as small as the capillaries we met in **B** for **BLOOD**.
5. These tiny branches are called **alveoli** and the capillaries are in their walls. Oxygen passes from the air through the wall of the alveoli and combines with the red blood cells in the capillaries.
6. When you breathe out, the diaphragm relaxes, the ribs move downwards and the air is squeezed out of the lungs along the same route it used to enter.

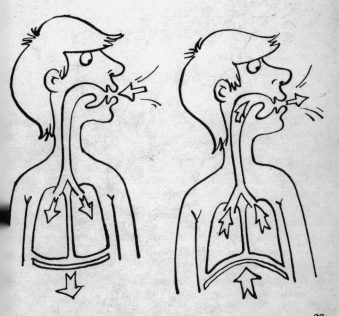

You may think it must be your nose or mouth that starts the breathing process, but as you can see it isn't. Breathing begins with the all-important diaphragm. In the past, when ladies thought they were being fashionable by squeezing their waists into corsets, they didn't realize that the tight corset prevented their diaphragms from working properly. This prevented them from breathing properly, with the result that all too often they fainted!

Sitting quietly and breathing normally, the average adult breathes about 7 quarts of air every minute.

Breathing through your nose like this, the wind speed in your nose is 4 miles per hour, which is the same speed as a gentle breeze.

When you are involved in active physical work, the amount of air you breathe goes up to something like 100 quarts a minute.

This means the wind speed in your nose is as fast as a really strong wind.

And, believe it or not, when you sneeze you can produce wind speeds as great as those in a hurricane or even a tornado!

A normal adult, sitting down reading a book or watching TV breathes in and out between 10 and 14 times a minute. Each breath lasts from 4 to 6 seconds. How many breaths do you take in a minute when you are sitting down? And how many do you take when you are walking and running?

Make a chart showing the different activities you are involved in every day and mark the number of breaths-per-minute by each one.

EAR! EAR!

As everybody knows, if you didn't have ears you wouldn't be able to hear. But did you also know that if you didn't have ears you wouldn't be able to stand up? Hearing is the obvious job that the ears do for us. But they do an even more important and basic job as well—they help us keep our balance. They give us our **equilibrium**. Inside the **inner ear** there are three **semi-circular canals** filled with liquid and two containers filled with tiny hairs and chalky substances. The canals and the containers, the fluid and the chalk, are the **vestibular apparatus**. Without them you wouldn't know whether you were coming or going, standing or falling—you'd really be living in what seemed like a topsy-turvy world.

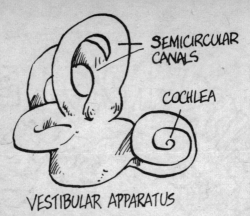

VESTIBULAR APPARATUS

Giving us our balance is the all-important job of the ears. Th[at] doesn't mean that giving us our hearing isn't an important job, to[o]. It is a vital job, otherwise everyone would be deaf.

The part of the ear on the outside of the head helps direct t[he] sound waves into the ear, but that's only a small part of it. The re[al] job of hearing is done inside. Sound waves entering the ear stri[ke] the eardrum, or **tympanic membrane**, causing it to vibrate. Touc[h]ing the inner surface of the eardrum is a tiny bone, the **hamme[r]**. The hammer connects by a joint to another little bone, the **anv[il]**. The anvil is jointed to a third bone, the **stirrup**. Below the stirrup [is] a coiled-up tube called the **cochlea**.

When sound waves make the eardrum vibrate, the eardrum mak[es] the hammer vibrate as well. The vibrating hammer strikes again[st] the anvil, which passes the vibration on to the stirrup. The stirr[up] makes the liquid in the cochlea vibrate and the vibration in t[he] cochlea sets up impulses in the nerves that travel to the brain. T[he] part of the brain that deals with hearing understands those impuls[es] as sounds. And you have two ears, both doing the same job, to he[lp] you find out where the sounds are coming from.

Sound waves are like invisible ripples in the air. You cannot s[ee] them, but you can feel them. The ripples travel at different speed[s]. These **sound frequencies** are measured in terms of **cycles per s[ec]ond**.

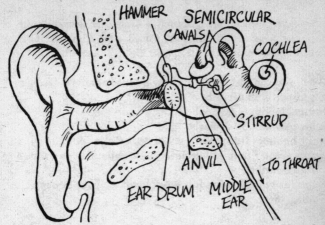

A good human ear can detect sound frequencies from 20 cycles per second to 20,000 cycles per second.

The sound of a man's voice is about 100 cycles per second.

The sound of a woman's voice is about 150 cycles per second.

The note middle C on a piano is 256 cycles per second.

20,000 cycles per second sounds just like a faint hiss to a human ear.

Dogs are among the few animals that can detect more than 20,000 cycles per second.

Sound frequencies are measured in cycles per second. Levels of noise are measured in **decibels**, named after Alexander Graham Bell, who also gave us the telephone.

When you can just, just, *just* hear a sound, that's 0 decibels.
A whisper is 30 decibels.
Normal talking is 60 decibels.
A loud shout is 90 decibels.
Noise becomes really nasty at 120 decibels.
Noises greater than 120 decibels cause pain and can damage your hearing.
From really loud noises it is possible to lose your hearing altogether.

Try these two experiments. The first will teach you about the balancing job done by your ears. The second has to do with hearing.

1. Spin around and around and around. As you spin around, the liquid in your ears is spinning around, too. If you suddenly stop, the liquid will go on spinning around for a few moments. The spinning liquid will send a message to your brain. You're now standing still, but the liquid goes right on spinning, making you very dizzy. (Don't get carried away with this experiment. Spinning so much that you fall over when you stop isn't good for you.)
2. Turn on the radio or put on a record. Now cover both your eyes and just one ear, but cover them well. Revolve a few times, then wander about the room. Try to decide where the sound is coming from. See if you're right. Repeat the experiment several times and you will be surprised by how often you choose the wrong direction. If you just cover your eyes, but listen for the noise with both ears, you will find the right direction every time.

FOOD FOR THOUGHT

We are what we eat—and if we didn't eat and drink we'd die. We need nourishment and we get it from our food. What's more, the right nourishment comes from a range of different foods that provide us with the different things we need.

We need **Carbohydrates**. Bread, spaghetti, sugar and starch are all carbohydrates. Carbohydrates are a source of energy, though if we have too many carbohydrates they turn into fat which our body stores to make us fat.

We need **Fat**. Butter, margarine, lard and olive oil are all fats. Fats are another source of energy and, on the whole, a better source of energy than carbohydrates. Again, if we have too much fat the body will store it and it will make us fat.

We need **Protein**. Protein is manufactured in the bodies of green plants, so when we eat green plants, and animals eat green plants, the plant protein is changed inside us and inside the animals into muscle. When we eat meat, which is the muscle of animals, we also get protein through the meat.

We need **Vitamins**. **Vitamin A** is important for general growth and for healthy eyes and skin. **Vitamin B** is important for general growth, healthy nerves, strong muscles and for good digestion. **Vitamin C** is important for general growth, for good teeth and skin, and for fighting germs. **Vitamin D** is important for strong bones and teeth. **Vitamin K** is important for the liver and for making the blood clot.

We need **Minerals**. We only need them in small quantities, but we need a wide range of them. Two of the most important are calcium and phosphorus for healthy teeth and bones.

To be healthy and to grow properly we need a **balanced diet**, combining all the different things we need in the right proportions.

Almost certainly somebody sometime has told you that you shouldn't eat too many sweets! Almost certainly they were right, because too many sweets, too much chocolate and too many rich cakes, are **bad** for you! But in a week when you are being sensible and not eating too many sweet things, what exactly do you eat? And

is it a healthful diet, or one that's **bad** for you? Keep a detailed **Diet Diary** for a whole week, making a note of everything—yes, everything—you eat.

At the end of the week, take each ingredient on the list—carbohydrates, fats, proteins, vitamins, minerals—and work out what you got from it. In that way find out how balanced your diet is. Your mother's cookbook will have a section on all these food groups.

Compare your diet with the diet of a friend and with the diets of adults you know, such as teachers, parents and grown-up brothers and sisters.

GERM WARFARE

Throughout your life your body is at war, fighting off the **germs** that are constantly attacking it. Once the germs are inside your blood stream, it is the job of the white corpuscles to trap the germs and destroy them. But the germs can be destroyed outside your body as well.

Sunshine kills germs.
Fresh air kills germs.
Soapy water kills germs.
Tears kill germs. Whenever you blink, tears wash your eyes.
Clean sweat kills germs—but sweat also traps dirt, which is why we wash it off.

Your outer ear contains wax and hairs to trap germs.
Your nose contains hairs to trap germs.

Most germs get into the body through your mouth.

Once they are inside your system, it becomes the job of the white corpuscles to lead the battle against them.

In your life so far your body has had to fight millions of different germs. You could never list them all. But what you can do is list all the **illnesses** from which you have ever suffered. Name each illness, and the year when you had it. Then, if you or your parents remember, find out what **caused** the illness, how **bad you felt** while you had it, and the **reason** you were eventually cured. A regular dictionary or children's encyclopedia can help.

HAIR RAISING

The average adult scalp contains 100,000 hairs.
Redheads have fewer hairs on their heads—around 90,000—and blondes have a lot more—up to 150,000.
A hair on your head grows nearly 8 inches a year.
It would take you about six years to grow a head of hair long enough to sit on!

Ever wondered why your hair doesn't hurt when it's cut? Because hair cells aren't alive. Hair is pushed up from a live root, the hair **follicle**, beneath the skin. Every few years the follicle has a rest and the hair growing out of it falls out. The rest period lasts some three to four months, after which the follicle wakes up and a new hair grows out. The daily loss of hairs from a full head of hair is between 30 to 60 individual hairs. Although we lose hairs every day, the follicles that grew the hairs aren't lost. They are just taking a rest and will be back in action before long. It is only with baldness that the follicles do not come back to life again and grow fresh hairs.

Our skin is covered with hair, but obviously it is the hair on our heads and, in the case of men, the hair on our faces that grows most quickly. In theory you could grow almost 40 feet of head hair in a lifetime, but because of the regular rest periods of each hair follicle every few years, it doesn't work out that way. Since the cycle of a hair follicle lasts from one to six years, the maximum length of a hair is likely to be around 3 feet. Longer hairs have been known, but not **much** longer ones!

Human hair may be dead, but it is still pretty strong.
A single hair can support a weight of 3 ounces.
The combined strength of all the hairs on a man's head will support 100 adults!
(Please take note: if you were to try supporting a hundred friends by the hairs of your head, the hairs would be pulled out at the roots. The hairs are strong enough for the job, but the roots aren't.)

Are you going to be bald? Well, if you're a girl, probably not—though women can and do go bald sometimes. If you are a boy and you want to know whether or not you are likely to go bald, look at your father's head and your grandfather's and, if you can get hold of the family photograph albums, look at your great-grandfathers' heads and their fathers' heads too. If you have a bald father and grandfather then the chances are that, one day, you too will go bald. Never mind—as Kojak once said, "Bald is beautiful!"

INTO THE INTESTINES

What happens to the food you eat? You know where it goes in and you know where it comes out, but what happens to it in between?

1. When you take a mouthful of food, you begin by chewing it with your teeth. In your mouth you add saliva to it and the saliva begins the **digestive process**.
2. The food, munched up and mixed with saliva, is now swallowed and passes into the food pipe called the **esophagus**. This is a muscular tube that pushes the food down into the **stomach**.

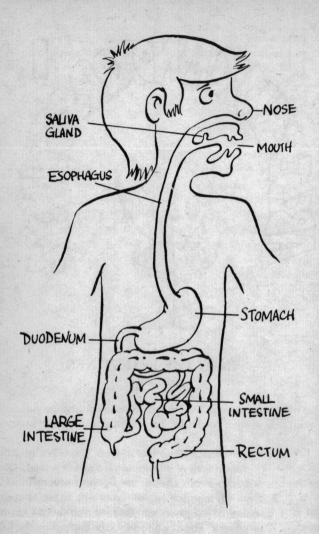

3. The stomach is coated with muscles. It is higher up in your body than you might suppose, just below your breastbone. In the stomach the food is churned around while digestive juices pour in from the stomach wall.
4. From your stomach the food moves on into the **small intestine**. This is where most of the digestive process takes place. The small intestine has a multitude of tiny projections on it called **villi**. The digested food, which is just a liquid by now, is absorbed through the villi and passes into the capillaries that are inside the villi. Now the nourishment from the food is inside the bloodstream.
5. The remaining food passes from the small intestine into the **large intestine**. Water is taken out of the food here and is absorbed into the bloodstream. The food that cannot be digested moves on to the lower part of the large intestine, the **rectum**. From there it is removed through the **anus** when you go to the bathroom.

You can swallow food standing on your head! Don't try, because it's not all that easy or all that good for you, but it's a fact. Food doesn't have to fall **down** into your stomach. It can travel **up** to your stomach, too. There are many other animals and birds who can eat and drink with their heads down so that the food and drink has to travel **up** to their stomachs.

Take a look at the way the animals around you—cats, birds, and dogs—eat. Notice which ones eat with their heads down, and which ones have to fling their heads back to swallow.

JOINTS AND BONES

Every **bone** in your body (except for one) meets up with another one. And the places where your bones meet and link up are called **joints**. Most of the joints are the movable ones that help you to do everything from getting up and running to bending down to touch your toes and wiggling them.

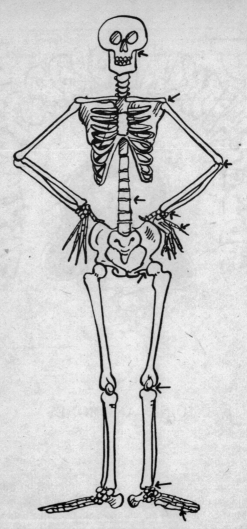

On this skeleton most of your important movable joints ha arrows pointing to them.

If you want to see how important your movable joints are, try keeping them all stiff for five minutes. You will find it difficult to move at all!

Not all the joints can move. In your **skull**, for example, there are eight bones joined together that make up the part of the skull that encases your brain. It is called your **cranium** and the bones in your cranium don't move.

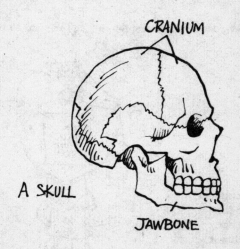

A baby is born with about 350 bones.
As you grow older lots of these little bones grow together to form larger, single bones.
The average adult has 206 bones in his/her body.
Some have one or two more, because their bones didn't grow together properly.
Others have one or two fewer, because the growing together system went a bit too far!

As your bones grow, you get taller, of course. Your growth in height is likely to stop by the time you are 16¼ if you are a girl, and 17¾ if you are a boy.

Between your 18th and your 30th birthdays, you may grow another ⅛ inch, but that's about it.

Let's take your skeleton from top to toe. There are 29 bones in your **skull**:

8 in the cranium
14 in your face
3 in each ear
1 in the throat

It's the one in the throat, the **hyoid** bone, that doesn't meet up with any other bone. **It's the only jointless bone in your body.**

The spine has 26 bones.
The chest has 26, of which one is the breastbone and 25 are ribs.
You have 2 collarbones.
You have 2 shoulder bones.
Each arm contains 1 upper-arm bone and 2 lower-arm bones.

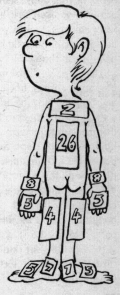

here are 8 bones in your wrist.
our palm has 5 bones, and there are 14 bones in the fingers of each hand.
ou have 2 hip bones.
ach leg contains a thighbone, a kneecap, a shinbone, and a bone on the other side of the lower leg.
here are 7 bones in the ankle of each foot.
bones are in the foot itself, with 14 bones in the toes of each foot.

51

If ever you are in bed and can't get to sleep, try moving all the bones in your body, starting with your toes. And if you ever want to prove to yourself that human beings have more useful bones than other animals, look at your hands and at the amazing things you can do with them. For example, try picking up a pencil or a cup **without using your thumb**. It isn't easy.

The bones at all the movable joints in your body are held together by tough, stringy tissue called **ligaments**. To help oil your joints and keep the movable pieces moving, a number of the bones contain a small hollow that produces a special lubricating liquid. The bones that meet up at joints that don't move are held together by a different kind of tough tissue called **cartilage**. Cartilage is what joins together the all-important bones of your spinal column and because it's springy it is a fairly successful shock absorber.

If you didn't have a skeleton inside you, your body would just collapse onto the floor like a wobbly rag doll. It's the bones of your body that give it its **shape**. The bones also help **protect** the soft parts of the body—the cranium protects the brain, the bony sockets at the front of your skull protect the eyes, the spinal column forms a tube of bone that protects your spinal cord, and the ribs protect your heart and lungs.

Your bones provide shape, support and protection. They also act as anchor points for your muscles. What's more, the yellowish marrow inside your bones makes all the red blood cells. Your bones also help destroy old red blood cells, produce other ingredients for your blood and contain important minerals—like calcium, phosphorus, magnesium and iron—that keep you healthy.

So without any bones, you wouldn't just look like jelly, you wouldn't be able to live at all!

1. Bones protect pieces of your body **inside** your body. What would you look like if your skeleton was on the outside like a bony suit of armor? Design a skeleton to wear on top of the rest of your body—and then work out why the system we have now is so much more practical.

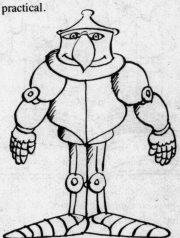

2. Of the 206 bones in the average adult body, each one has a name, and each name is something of a mouthful. For example, a medical student could tell you that the 31 bones in each of your legs have some splendidly tongue-twisting names. There's 1 **hip bone** (that's easy), 1 **femur** (a bit trickier), 1 **kneecap** (easy again), 1 **tibia** and 1 **fibula** (trickier again), 7 **tarsals**, 5 **metatarsals**, and 14 **phalanges**!

Find a medical textbook that gives you the names of all 206 bones in the body. See if you can learn the names of the bones in one part of the body—such as your hand—by heart.

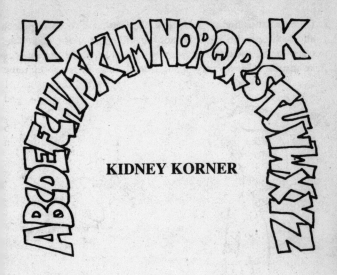

KIDNEY KORNER

Human beings have two **kidneys**. They are about 4½ inches long, 2½ inches wide, 1½ inches thick and weigh about 5 ounces. They have two jobs:

1. Blood goes into the kidneys to be cleaned. Clean blood goes back into the blood stream from the kidneys. Waste comes out of the body (and into the toilet) as urine.
2. The kidneys regulate the amount of salt and liquid in the body.

If you want to find your kidneys, they are close to your spine in front of the twelfth ribs. While you are looking for them (and amazed at the thought that over two pints of blood are pumped through your kidneys every minute), think about the name **kidneys**.

Kidneys is a very old word. It's been around for at least 1,000 years, though when people first talked about kidneys they didn't know what part kidneys played in our bodies. It's a very good word. But can you think of a better word for kidneys? How about **kleenbl** or **bublubu** or **pom**? Take all the main parts of the body and think of your own names for them. Start with everyday words—like **head**, **neck**, **shoulder**, **chest**, **heart**, **legs** and so on—and then move on to more complicated words—capillaries, intestines, plasma, ligaments—and see if you can invent names for different parts of the body that sound even better than the names they already have.

LEFT-HANDED

Are you left-handed? If you are, you may like to know that there are roughly 200,000,000 other people like you in the world today. That's an awful lot of people, but it still amounts to only about 5 percent of the world's population.

Of the 19 out of every 20 people who are right-handed almost all of them were born that way. Some were forced to use their right hand instead of their left (which is not a good thing), and others had to learn to use their left hand when they lost the use of their right.

(An example of this is Mickey Lolich, the baseball pitcher, who had to learn to throw with his left arm after injuring his right arm in a bicycle accident.)

Way back in the Stone Age the number of left-handed people was about equal to the number of right-handed people.

By the Bronze Age, the number of right-handed people had greatly increased—only a quarter were left-handed then.

What is the explanation? Nobody really knows.
Some scientists have suggested that right-handedness increased when warriors began to hold their shields in their left hands to protect their hearts and used their right hands to hold swords or spears.
Can you think of any other explanation?

For an hour or a morning or even a whole day, reverse the work of your hands. Everything you normally do with your right hand make your left hand do. And everything you normally do with your left hand, make your right hand do.

MUSCLE POWER

To move you need **muscles**. Your bones couldn't move without muscles to move them. In your body there are about three times as many muscles as there are bones.

> Inside your body there are roughly 656 muscles and they are heavy.
> About 42 percent of a man's weight is muscle.
> About 36 percent of a woman's weight is muscle.

Muscles are made of bunches of muscle tissue held tightly together. Most of the muscles attached to your skeleton are linked to a bone at one end by a **tendon** or (much less often) at both ends. Your tendons can be very small or as long as 12 inches.

The muscles that move your skeleton are called **voluntary muscles** because you can move them at will. But there are other muscles in your body, the **involuntary muscles**, that you can't move at will. In your eyes you will find a good example of the difference between voluntary and involuntary muscles. The voluntary muscles allow you to control your eyes to look wherever you want to look. But you cannot control the involuntary muscle that widens and narrows your pupils.

Muscles can only work by pulling. Even if you are pushing something as hard as you can, your muscles are in fact doing the job by pulling. By shortening and lengthening your muscles, you can make them bend or straighten your joints.

Feel some of your muscles at work:

1. Rest your right elbow on a table. With your left hand feel the middle or the top part of your right arm. Raise your right hand slowly, with your elbow still on the table, and you can feel the muscle in your right arm (the **biceps**) swelling up as you shorten it.

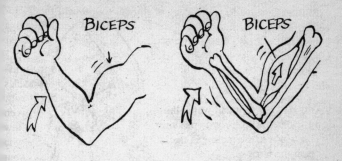

2. Put your left hand around your right forearm and drum your right-hand fingers on the table. You can now feel your forearm muscles at work.

3. Waggle your foot and watch the muscle bulge up toward your knee. That's where the muscle is that's doing the work.

Now feel two of your tendons:

1. Bend your arm and feel the tendon inside your elbow.

2. The big tendon here steadies the whole weight of your body when you stand on one foot.

Feel how hard it is.

Take a close look at a piece of muscle. You don't need to cut yourself up, just use a small piece of beef because beef is actually muscle.

With a pin, pick apart a piece of raw beef. Separate it into long, thin strands. These are the fibers of the muscle tissue.

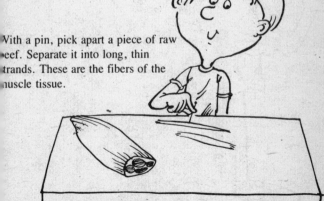

If you could look at a thin muscle fiber through a microscope, you would be able to see the muscle tissue made up of long, slender cells.

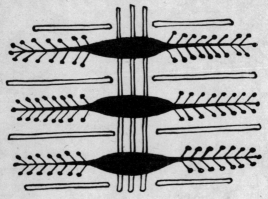

NOSE KNOWS

First joker: My dog has lost his nose, you know.
Second joker: How does he smell?
First joker: Terrible!

If you didn't have a nose you might still stink, but you couldn't smell. There is a small patch of special cells on the upper part of the inner surface of your nose. When the gases that float away from the things that have a smell come into your nose and reach that small patch of cells, the cells cause impulses to travel along the nerves to your brain. That's when you smell the smell.

Your smelling organ is more sensitive to some smells than to others. When you have a bad cold, you may lose your sense of smell because the mucus that forms in your nose when you have a cold covers the patch of sense cells and prevents the gases giving off the smell from coming into contact with them.

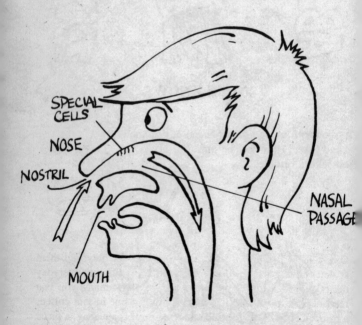

As well as helping you to smell, your nose cleans and warms the air you breathe. You breathe air in through your **nostrils**, two tunnels that lead to a small cave inside your face. The air we breathe is full of germs and specks of dirt and the hairs inside your nose help to clean the air. Germs and dirt get trapped in the hairs. The sticky lining to your nose helps catch the germs and dirt as well. Tiny blood vessels in your nose also warm the cold air before it travels down your windpipe to your lungs.

Give yourself a Smell Test. Close your eyes (or blindfold them securely) and get a friend to bring you lots of different everyday things that have different smells and see how many you can recognize. Without looking, sniff at a rose, some instant coffee, a piece of chalk, strawberry jam, a cauliflower, a teabag, a piece of wood, a jar of paint. Award yourself one point for every smell you get right.

2. When you have finished your Smell Test, you may think that your sense of smell is good. For a human being, it's probably very good. But by a dog's standards, your sense of smell is very, very poor indeed.

> Over all, a dog's sense of smell is about a million times better than a man's!
> If a dog knows his master, that dog will be able to recognize the master's smell on an old stick that the master has held in his hands for only a few seconds.
> It doesn't matter how many people have handled the stick before or after the master. The dog will still know his own master's smell.

If you have a dog, give it some tests and see if it can recognize your smell.

OPTIC ORGANS

No, optic organs aren't musical instruments. They're your eyes, the wonderful parts of the body that enable you to see the world around you—and read the words you are reading now.

> A fully grown human eyeball weighs about ¼ ounce and has a diameter of slightly less than 1 inch.
> A man's eye is a fraction bigger than a woman's—though it isn't as big as an eagle's eye.
> Birds have very large eyes for the size of their bodies. An owl's eyes, for example, take up one-third of its head.

When a baby is born the size of its eyes is much nearer the size they will be when the baby has grown up than the rest of its body. Between being born and becoming an adult, your body increases its volume 20 times; your eye only increases its volume 3¼ times.

A baby's eyes are one four-hundredth of its weight, but an adult's eyes are one four-thousandth of its weight.

If you could cut across one of your eyes and look inside it, this is what you would find:

DIAGRAM OF THE EYE

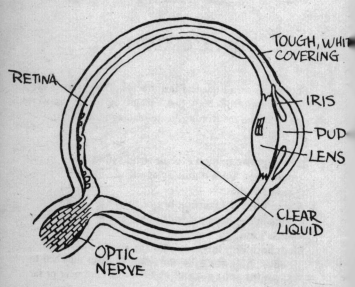

Your eye is surrounded by a tough, white protective covering. At the front of your eye, part of this covering is round and see-through. Behind this see-through part of the eye is a space filled with clear liquid. At the back of this space is a circular tissue called the **iris** with a hole in it called the **pupil**.

The iris is the colored part of your eye. On the inner edge of it, around the pupil, you will find a ring of little muscles, all of them sensitive to light. In a bright light the muscles *tighten* and the pupil gets smaller. In a low light the muscles *relax* and the pupil gets bigger.

Behind the iris is a round see-through **lens**. Muscles attached to the rim of the lens can make it focus on things that are near or far away. You are able to see things because light rays bounce off objects and travel into your eyes. In the pitch dark you can't see because there are no light rays reaching your eyes.

When a light ray passes through the lens in your eye, it is turned **upside down** and **reversed from right to left**. After going through the lens, the light travels through a small cave filled with clear liquid. All over the inside surface of this cave are nerve ends that are sensitive to light. The surface with the nerve endings in it is called the **retina**. The nerve endings link up with the **optic nerve** and the optic nerve takes the picture up to your brain. Your brain then turns the picture the right way up and the right way around so that you see it properly.

Each of your eyes sees a slightly *different* picture of the world. Look at an object with both eyes, then close one eye. Then close the other eye (opening the first eye again) and you will see that the picture you get of the object looks slightly different each time.

Your brain joins together the pictures you get from your left and right eyes. With your two eyes working together you see better and you are able to work out how far or how near objects are. Put an object on a table, sit at the table and close one eye. Now try and touch the object. Do it 10 times. You will probably miss the object a few times. With both eyes open you should be able to touch the object every time.

1. Watch your pupil shrink. Go into a very brightly lit room and stand in front of a mirror. Now cover up one eye completely for at least 2 whole minutes. Now uncover the eye and watch the pupil shrink! By getting smaller in a very bright light, the pupil protects your eye. By getting larger in dim light it helps you to see better.

2. Find out how you see better with two eyes. Stretch out your arms.

Now bring your arms together so that the tips of your two outstretched forefingers meet, like this:

Do it with both eyes open and you will find that making your fingers meet isn't difficult. Do it with one eye shut and you will find it a lot harder.

3. Find your **Blind Spot**. At the point where the optic nerve comes into your eye there is no retina, so this tiny area in your eye is not sensitive to light. This tiny area, which is just below the middle of the back of your eye, is your Blind Spot.

To prove to yourself that you've got one, try this experiment. Look at these two letters:

A **B**

Close your left eye and hold up the page in front of your right eye. Look at the **A**. Now move the page toward you and away from you until you find the point where the letter **B** disappears! At this point the **B** is focused by the lens of your eye exactly on your retina's Blind Spot.

PERCEPTION

Perception is turning what we see into something we understand. Our perception is *our view of the world*. Look at this drawing, and what do you perceive?

A duck? Or a rabbit?

You could perceive that picture in two different ways. It could be a duck **or** a rabbit. Generally, our perception of the world improves as we grow older and we learn to recognize more things for what they are. For example, a 5-year-old looking at these two rows of dots will agree that there are the same number of dots in row **A** as there are in row **B**:

But if you showed the 5-year-old these two rows of dots, he'd tell you that there were **more** dots in row **B** this time:

By the time you are 8 or so you wouldn't make that mistake. You would be able to perceive that there were the **same** number of dots in rows **A** and **B** but the dots in row **B** were spaced out more.

Look at all the dots on the next page. They are all equally spaced, but they do seem to form patterns, don't they? The brain tends to group a mass of dots like these into different patterns. That's how we perceive them.

Sometimes we are the victims of **optical illusions**, meaning that we perceive something to be so that isn't actually so. Our perception is fooling us!

See if you can answer these questions correctly.

1. In which of the two pictures is the moon bigger?

2. Of these three cubes, which one is nearest to us and which one is the largest?

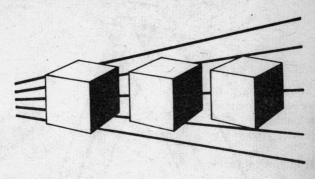

3. Are any of these lines straight?

4. What's wrong with this triangle?

ANSWERS

1. The moon is the same size in both pictures. It just seems bigger in the second picture because it is nearer the ground, so we feel it **must be** bigger.

2. All three cubes are the same size, as you can now see. The background gives the impression that the third cube is larger and farther away.

3. All the lines are straight.

4. The triangle is an **impossible** object. It can be drawn, but it can't be made. Look at the picture again closely and see if you can work out **why** you couldn't make the triangle out of wood.

QUICK THINKING

The human **brain** isn't the biggest in the animal kingdom—it's a quarter the size of an elephant's, for example—but it is the best. Without it, you'd be at a loss for words, and everything else as well!

> The brain is a soft lump of about 14,000,000,000 cells.
> It weighs just over 3 pounds in an adult and occupies the upper half of the skull.

The largest part of the brain is the **cerebrum**. It controls your thinking in all its many aspects. At the back of your skull is the **cerebellum**. It coordinates your muscular activity. The cerebellum controls all the voluntary muscles automatically. The involuntary muscles are controlled by a part of the brain called the **medulla**, which is just at the top of the **spinal cord**.

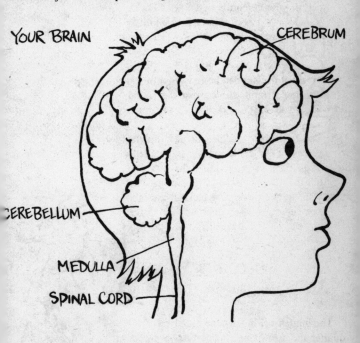

The spinal cord goes down from the medulla through four-fifths of the length of your spine. Nerves branch off the spinal cord and carry orders from your brain to all parts of your body and carry messages from your senses to your brain.

For example, if you drop a cup and it breaks on the floor and you want to pick up the pieces, this is what happens:

1. When the sound of the cup hitting the ground reaches your ears, a message goes along the two auditory nerves to your brain.

2. When the messages reach your brain, part of your cerebrum perceives them as sound, and passes this information to another part of the cerebrum which recognizes the sound.

3. This part of your cerebrum is your memory and it reminds you that the sound you've heard is of a cup hitting the floor.

4. The brain now decides to pick up the pieces. Your cerebrum sends hundreds of different messages along the nerves to tell the many muscles involved that you want to bend over and pick up the broken cup.

Your brain is the control room of your body. Different parts have different jobs to do.

The **cerebrum** receives the messages and stores them.

The **motor cortex** of the cerebrum controls the muscles and sends out signals.

The **cerebellum** coordinates, balances and sends the orders to the muscles you can control.

The **medulla** controls the muscles you can't control—your heartbeat, your rate of breathing, the movements of your stomach and intestines, and many others.

Instead of drawing your own brain as a brain, draw a picture of it as a control room.

Here's Rowan Barnes Murphy's picture of his brain at work:

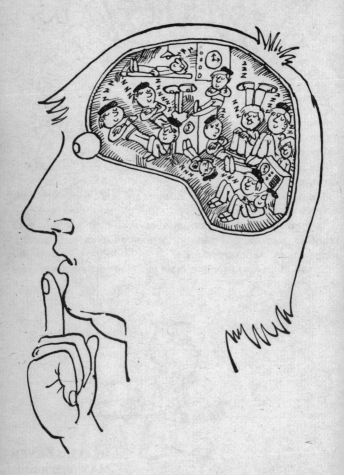

What does yours look like?

HEAVIEST-EVER WOMAN: Pearl Washington—880 lb.

HEAVIEST-EVER MAN: Robert Earl Hughes—1,069 lb.

TALLEST-EVER WOMAN: Jane Bunford—7 ft 11 in.

TALLEST-EVER MAN: Robert Wadlow—8 ft 11 in.

SHORTEST-EVER MAN: Calvin Phillips—26.5 in.

SHORTEST-EVER WOMAN: Pauline Musters—23.2 in.

Weigh and measure yourself and all your friends. Find out who is the tallest and shortest. Who is the lightest and heaviest? And keep a record of how you grow taller and heavier as you grow older.

SKIN TIGHT

The average adult is covered with about 18 square feet of skin!

Your skin is—
Airtight and waterproof.
It protects your body from germs and from drying up.
It helps to keep the temperature of your body even. When it's very cold, the blood vessels in your skin contract and force the blood deeper into your body. When it's very hot, the same blood vessels expand and bring more blood to the surface of the skin.

There are two layers to your skin. The top layer is the **epidermis**. It is made of dead, flattened cells that are falling off your skin all the time. Below the epidermis is the second layer, the **dermis**, which is made of living cells. When the dead cells of the epidermis fall off the skin, cells from the dermis move up to the epidermis and replace them.

In the dermis are thousands of blood vessels and nerve endings, which is how you can *feel* things through your skin. You also *sweat* through your skin. Small tubes in the dermis open through the epidermis. They are your **sweat glands** and the openings are called **pores**. When it is hot, the sweat glands make more sweat and as the sweat dries it cools down your skin.

When you are hot, more blood moves through the vessels near the surface of your skin so that the air can cool it. Color cells in the

deeper part of the epidermis give your skin its color. Sunlight can make those color cells grow darker and the darker skin can protect your body from some of the sun's harmful rays.

When it is cold your surface layer of blood vessels shuts off and the deeper ones open up to help your skin keep in the warmth. This makes your skin look paler. The cold can make your skin muscles tighten. This releases a special oil that makes the hairs on your body stand up and makes **goose pimples** on your skin.

Build up a fingerprint collection. The pattern of the skin of everybody's body is slightly different. Take all your friends' fingerprints by pressing them into a stamp pad (with **washable ink**), and study the differences.

TEETH TIME

Here's a mouth full of **teeth**:

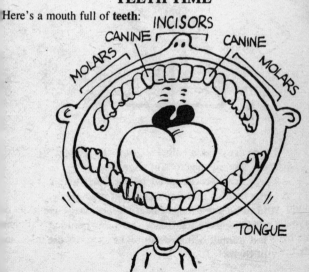

There are thirty-two teeth in the mouth of an adult. Everyone has two sets of teeth in his lifetime.

In the first set, which you grow as a baby and as a young child, you have just 20 teeth.

In the second set, which you grow as you get older and the first set of teeth falls out, you have 32. (Some people, who lose their second set of teeth, have a third set. But they don't grow that set—they have it made for them by the dentist!)

Look at the mouth full of teeth again.

There are 8 teeth called *incisors*. They chop the food at the front of your mouth.

There are 20 teeth called *molars*. They grind the food so that you can digest it properly.

And there are 4 fangs called *canines*. Dogs use their fangs to tear meat and gnaw bones, but human beings don't make as much use of theirs.

Teeth go bad because liquid from chewed food sticks to them and makes holes in them. The germs in the air make the holes deeper and they become **cavities**. The dentist fills the cavities with **fillings** so that germs can't live in the cavities. If germs are allowed to live in the cavities, your teeth will be ruined and will fall out.

That's why cleaning your teeth and gums regularly is **very** important.

PROJECT!

When you clean your teeth you almost certainly use a toothbrush and toothpaste. But there are millions of people around the world who don't use a toothbrush or toothpaste. So how do they clean their teeth? Try to find out the different ways in which people—particularly in Africa, South America, India and the Far East—clean their teeth. (Your dentist probably knows some ways.)

UGH!

"Ugh!" you say when you taste something you don't like. But how do you know that you don't like the taste? Well, your **taste buds** tell you.

Your taste buds are just below the surface of your tongue and in three different places in your throat. There are four tastes that you can detect through your taste buds—**salty**, **sweet**, **sour** and **bitter**. You detect the different tastes in different places.

The taste buds at the *side* and *tip* of your tongue detect saltiness. The buds at the *tip* of your tongue detect sweetness, too. Those at the *side* detect sourness and the ones at the *back* detect bitterness—which is why you often don't notice a bitter taste until you are about to swallow and it's almost too late to spit it out! You have about 3,000 taste buds in all, but almost none of them are in the middle of your tongue.

Sweet, sour, salty and bitter are the only things you can taste through taste buds. You think you can taste lots of other things as well, but you can't. You **smell** them. When it comes to recognizing the flavors of food, your sense of smell plays the most important part. Without a sense of smell, you wouldn't even be able to taste an onion! And when you have a cold and can't smell, the only tastes you will recognize are the four your taste buds detect for you.

Nobody knows exactly how the taste buds work. Scientists are investigating how our senses of taste and smell operate, but they still have a lot to learn. Since trained scientists still have not discovered all the secrets of the taste buds, you would find working out how the taste buds work a very, very big project—the work of a lifetime, probably.

If you can't spare a lifetime, play the **Taste Buds Game** instead. Blindfold your friends and give each of them different things to taste. Give **salty** things to one (salty peanuts, for example), **sweet** things to another (sugar, for example), **sour** things to another (lemon juice, for example) and **bitter** things to another (orange peel, for example).

When they have each recognized the tastes you gave them, get them to tell you the position of the taste buds that detected the taste for them.

VOCABULARY

Doctors and scientists use all sorts of complicated words to describe different parts of the body. Here is a list of all the unusual words used in this book. If there are any of the words in the list that you don't recognize, or that you recognize but can't remember what part they play in your body, see the letter by the word. Turn to that letter in this book and find out about the word.

Alveoli	D	Balanced diet	F
Anus	I	Biceps	M
Anvil	E	Blind Spot	O
Arms	J	Blood	B
Artery	B	Bone	J
Auricles	B	Brain	Q

Canines .... T	Iris .... O
Carbohydrates .... F	
Cartilage .... J	Joints .... J
Cavities .... T	
Cells .... A	Kidney .... K
Cerebellum .... Q	
Cerebrum .... Q	Large intestine I
Chest .... J	Legs .... J
Collar bones .... J	Lens .... O
Cranium .... J	Ligaments .... J
	Lungs .... D
Dermis .... S	
Diaphragm .... D	Medulla .... Q
Digestive process I	Minerals .... F
	Molars .... T
Epidermis .... S	Motor nerve .... Y
Equilibrium .... E	Muscles .... M
Esophagus I	
Eye .... O	Nose .... N
	Nostrils .... N
Fat .... F	
Fillings .... T	Optical illusions .... P
Follicle .... H	Optic nerve .... O
	Organ .... A
Germs .... G	Oxygen .... D
Hair .... H	Perception .... P
Hammer .... E	Pitch C
Heart .... B	Plasma .... B
Hemoglobin .... B	Platelets .... B
Hemophilia .... B	Pores .... S
Hip bones .... J	Protein .... F
	Pupil .... O
Incisors .... T	
Inner ear .... E	Rectum I
Involuntary muscles ..... M	Red cells .... B

103

Reflex action	Y
Retina	O
Senses	Y
Sensory nerves	Y
Shoulder bones	J
Skull	J
Sleep	Z
Small intestine	I
Sound frequencies	E
Spinal column	J
Spinal cord	Y
Spine	J
Stirrup	E
Stomach	I
Sweat glands	S
System	A
Taste buds	U
Teeth	T
Tendon	M
Tissue	A
Trachea	D
Tympanic membrane	E
Valve	B
Vein	B
Ventricles	B
Vestibular apparatus	E
Villi	I
Vitamins	F
Vocal cords	C
Voluntary muscles	M
White cells	B
Windpipe	C
X-ray	X

Write your own **Dictionary of the Human Body**. Take each one of the words in the list above and write a short description of what the word means and what it does. And if you come across other words that aren't in the list that you don't understand, find out what they mean, too.

WHAT GOES WHERE?

WHAT WENT WHERE

The eating machinery.   **The breathing machinery.**

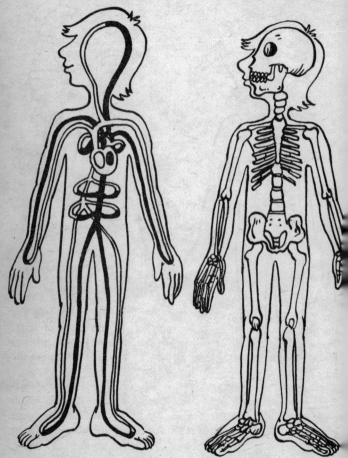

X-RAYS

X-rays are a form of radiation, like light and heat. They are invisible and they travel in straight lines at the speed of light, 186,000 miles per *second*. They can penetrate very thick material (up to 11 inches of solid steel) and have no difficulty penetrating the human body.

Although X-rays are invisible, if they are directed at a special screen the material on the screen will make the X-rays glow. If you put your foot in front of the screen, the shadow of your foot would be cast on the screen with the bones inside your foot casting an even deeper shadow.

X-rays are used by doctors to examine the insides of people (and, in some instances, to help them cure diseases). Can you imagine 25 other rays, like X-rays, but doing different work? Take each letter of the alphabet and invent a ray for it. You might even invent a ray that could help show us the way our taste buds work!

YOU'VE GOT A NERVE

Messages travel to and from your brain through your nerves. From your brain the main path they take is along the **spinal cord**. The main nerves connect with the spinal cord and tiny nerves join up with the bigger ones.

We are made aware of the world we live in by our **senses**. Five of them are ones that people have believed for hundreds of years were our only senses:

SIGHT
HEARING
TOUCH
SMELL
TASTE

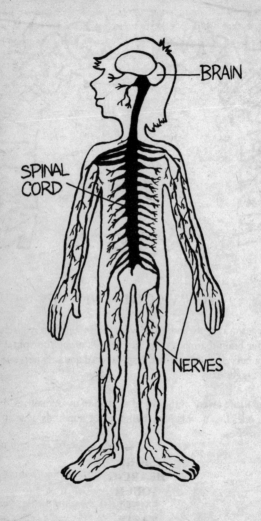

The nerves take the messages from the sense organs to the brain. The nerves that do this work are called **sensory nerves**. When you stub your toes, nerve endings in your toes send the message of pain up to your brain. Your brain **understands** that it is pain, but you actually **sense** the pain in your toes.

When the brain sends orders out to the body the nerves it uses to carry these orders are called the **motor nerves**.

Sometimes you react to something without having to think about it and send messages to the brain in the normal way.

For example, if you put your hand on a boiling kettle, you will pull it away at once. Your automatic action is called a **reflex action**.

With a reflex action, the messages take a special short-cut, so that you react automatically and immediately. The reflex action makes you draw your hand away from the hot kettle right away. At the same time, the message telling your brain to feel pain is traveling through the **sensory nerves** to the brain.

Test one of your reflexes just as a doctor does. Sit down and cross your left leg over your right leg like this:

Feel just below the kneecap of your left leg for the tendon that runs down from the kneecap. With the edge of your hand (or a ruler) strike the tendon sharply. Your left leg will jump upward. You can't stop it. You don't need to think about it. It's a **reflex action**.

Do it a few times and you may find that your leg begins to move **before** you actually feel the sensation of your hand (or the ruler) hitting your knee!

ZZZzzzzzz

To live at all you need air and food and drink and **sleep**. Believe it or not, you need sleep as much as you need food. A normal human being will die quicker without sleep than he will without food.

While elephants and dolphins can sleep for just 2 hours out of every 24, the average length of a night's sleep among adult human beings is 7 hours and 36 minutes. Children need to sleep much longer. People in their 50's need less sleep than people in their 20's, but people in their 60's get more sleep than at any time since childhood. On the whole, adult men sleep about 10 minutes longer than adult women.

PROJECT!

Make a **Sleep Chart**. Record the number of hours that you and each member of your family sleeps. Compare your family's Sleep Chart with the Sleep Charts of your friends. Good night. Sleep tight.

INDEX

alveoli, 23
artery, 13, 14
bacteria, 10-11
balance, 27, 31
baldness, 42
blind spot, 76
blood, 8-18, 55, 92, 108
bones, 47, 49-54, 108
brain, 84-88, 113-114
breathing, 22-26, 107
bronchial tube, 22
capillary arteries, 14
carbon dioxide, 9
cartilage, 52
cells, 7, 8, 84, 93-94
decibels, 30
diaphragm, 22-24
diet, 33-35
digestive system, 43-44
dizziness, 31
ears, 27-31
eating, 107
equilibrium, 27
eyes, 71-76
fat, 6, 32, 35
fingerprints, 94
food, 32-35, 43-46
germs, 36-38
goose pimples, 94
hair, 39-41
hands, 57-60
heart, 11-17, 108
hemoglobin, 10
intestines, 44-45
joints, 47-49
kidneys, 55-56
ligaments, 52

lungs, 13, 15
muscles, 61-66
nerves, 111-116
nose, 67-70
nostrils, 69
optical illusions, 80-83
organ, 7
oxygen, 9, 10, 22
perception, 77-83
plasma, 11
platelets, 9, 11
pulse, 16-17
red blood cells, 9, 10, 13, 53
reflexes, 111-116
senses, 111, 113
sensory nerves, 113, 114
skin, 92-94
skull, 49, 50
sleep, 117-119
smell, sense of, 68, 70, 100
sound frequencies, 28, 30
spinal cord, 85, 111
stomach, 43-46
sweat glands, 93
taste buds, 99-101
"Taste Buds Game," 101
teeth, 95-98
tendon, 62, 64-65
tissue, 7
vein, 13-15
vocal chords, 18-19, 20
water, 6
weight, 89, 91
white blood cells, 9, 10-11, 37
windpipe, 18-19
X-rays, 109-110